MATHS
Workbook

Level 2

MOONSTONE

Published in Moonstone
by Rupa Publications India Pvt. Ltd 2022
7/16, Ansari Road, Daryaganj
New Delhi 110002

Sales centres:
Allahabad Bengaluru Chennai
Hyderabad Jaipur Kathmandu
Kolkata Mumbai

ISBN: 978-93-5520-722-7

First impression 2022

10 9 8 7 6 5 4 3 2 1

The moral right of the authors has been asserted.

Printed in India

Contents

Numbers

1. Complete the number chain by counting forwards or backwards.

(a) 243 — ___ — ___ — ___ — ___ — 248 — ___

(b) ___ — ___ — ___ — 397 — ___ — ___ — ___

(c) ___ — 921 — ___ — ___ — ___ — ___ — 926

(d) 836 — ___ — ___ — ___ — ___ — ___ — 842

(e) ___ — ___ — ___ — 451 — ___ — ___ — ___

(f) ___ — ___ — ___ — ___ — ___ — 785 — ___

(g) ___ — 159 — ___ — ___ — ___ — ___ — ___

(h) ___ — ___ — ___ — ___ — ___ — 680 — ___

(i) ___ — 598 — ___ — ___ — ___ — ___ — 603

(j) ___ — ___ — ___ — ___ — ___ — 537 — ___

Comparing Numbers

1. **Fill with the symbols >, < or = to compare the numbers below.**

(a) 142 ☐ 135

(b) 382 ☐ 238

(c) 776 ☐ 589

(d) 124 ☐ 335

(e) 623 ☐ 436

(f) 109 ☐ 190

(g) 186 ☐ 97

(h) 928 ☐ 982

(i) 129 ☐ 271

(j) 635 ☐ 481

(k) 198 ☐ 231

(l) 444 ☐ 444

2. **Complete the table given below by identifying the smallest and the largest number in each group.**

	Smallest number	Largest number
(a) 136, 451, 286		
(b) 992, 299, 929		
(c) 460, 698, 632		
(d) 753, 729, 781		
(e) 875, 387, 178		
(f) 400, 600, 100		

Ordering Numbers

1. **Arrange the numbers of each group in the order that is asked for.**

(a) 452, 782, 389, 206, 193

	H	T	O
Smallest →			
Largest →			

(b) 647, 883, 527, 991, 400

	H	T	O
Largest →			
Smallest →			

(c) 708, 752, 731, 767, 795

	H	T	O
Largest →			
Smallest →			

(d) 296, 962, 975, 597, 695

	H	T	O
Smallest →			
Largest →			

(e) 97, 793, 379, 397, 739

	H	T	O
Largest →			
Smallest →			

(f) 429, 629, 829, 129, 529

	H	T	O
Smallest →			
Largest →			

Skip Counting

1. **Identify the rule of counting numbers for each group. Continue the counting to complete the patterns.**

(a)

220	222	224	___	___	___	___	___

(b)

541	546	551	___	___	___	___	___

(c)

860	863	866	___	___	___	___	___

(d)

340	350	360	___	___	___	___	___

(e)

791	794	797	___	___	___	___	___

(f)

958	963	968	___	___	___	___	___

(g)

196	296	396	___	___	___	___	___

(h)

673	675	677	___	___	___	___	___

(i)

485	490	495	___	___	___	___	___

(j)

136	140	144	___	___	___	___	___

Even and Odd

1. **Colour the shapes with the even numbers in them.**

470

389

293

972

364 + 103

42 + 38

361

175 + 121

2. **Read the numbers given below and group them into even and odd categories.**

427   364   548
962   239
394   425   76
163   869   51   840

Even numbers	Odd numbers

Number Names

1. Write the number names for the following numbers.

(a) 348 _____

(b) 673 _____

(c) 192 _____

(d) 885 _____

(e) 751 _____

(f) 206 _____

(g) 490 _____

(h) 646 _____

(i) 999 _____

(j) 527 _____

(k) 301 _____

(l) 1000 _____

2. Solve the given problems and write the number names of the answers.

(a) 39 + 48 = _____

_____

(b) 640 + 287 = _____

_____

(c) 399 + 420 = _____

_____

(d) 159 + 80 = _____

_____

(e) 12 + 10 = _____

_____

(f) 475 + 525 = _____

_____

(g) 829 + 43 = _____

_____

Place Value

1. **Write the place value of the underlined digit. One has been done for you.**

Number	Place value	Number	Place value
(a) 42<u>9</u>	9 ones	(h) <u>7</u>23	_____
(b) <u>7</u>86	_____	(i) 5<u>0</u>9	_____
(c) <u>9</u>05	_____	(j) 86<u>4</u>	_____
(d) 1<u>4</u>3	_____	(k) 9<u>3</u>1	_____
(e) <u>2</u>20	_____	(l) <u>1</u>57	_____
(f) 63<u>8</u>	_____	(m) <u>4</u>90	_____
(g) <u>1</u>000	_____	(n) 3<u>6</u>3	_____

2. **Read the clues below and write the numbers.**

(a) I have 3 at the hundreds place, 0 at the tens place and the smallest even number at the ones place.

(b) I have the smallest odd number at the thousands place and all other places have no value.

(c) All my places, hundreds, tens and ones have the same number which is more than 7 and less than 9.

(d) I have 9 at the ones place and 1 at the tens place. My hundreds place has an odd number which is more than 3 and less than 7.

(e) I have zero at both the tens place and the ones place. My hundreds place has an even number which is more than 5 and less than 8.

3. Count the base ten blocks and write the numbers.

(a)

(b)

(c)

(d)

(e)

12

Expanded Form

1. **Write the standard form of the numbers given in expanded form.**

   (a)   800 + 90 + 7   ⟶ ⬚

   (b)   1000 + 200 + 80 + 6   ⟶ ⬚

   (c)   700 + 9   ⟶ ⬚

   (d)   500 + 40 + 2   ⟶ ⬚

   (e)   300 + 60   ⟶ ⬚

   (f)   900 + 30 + 1   ⟶ ⬚

2. **Write the following numbers in their expanded form.**

   (a)   645   _____

   (b)   796   _____

   (c)   1043   _____

   (d)   518   _____

   (e)   410   _____

   (f)   907   _____

   (g)   212   _____

   (h)   380   _____

3. **Make the greatest and smallest numbers with the digits given below.**

Digits	(a) 4, 6, 3	(b) 7, 1, 9	(c) 3, 5, 8	(d) 2, 8, 6
Greatest number				
Smallest number				

4. **Compare the numbers using the signs > , < or =.**

(a) 740 + 9 ◯ 680 + 10

(b) 823 + 100 ◯ 442 + 200

(c) 600 – 30 ◯ 400 + 20

(d) 500 + 100 ◯ 900 – 300

(e) 360 + 40 ◯ 550 – 100

(f) 286 + 4 ◯ 230 + 40

(g) 142 + 8 ◯ 140 + 10

(h) 300 + 7 ◯ 286 + 48

(i) 150 + 250 ◯ 300 + 150

(j) 650 – 150 ◯ 750 – 400

(k) 800 – 120 ◯ 980 – 300

(l) 197 – 97 ◯ 98 + 10

Addition

1. Follow the steps to add the numbers and find their sum.

(a)
```
    4  2          +  2  4
 +  1  9
```

(b)
```
    3  6          +  4  0
 +  2  6
```

(c)
```
    2  8          +  3  5
 +  1  6
```

(d)
```
    1  0          +  2  8
 +  1  9
```

(e)
```
    3  3          +  4  4
 +  1  8
```

(f)
```
    2  9          +  5  1
 +  1  4
```

2. Add the three digit numbers given below.

(a)
```
    2  4  4
 +  6  3  9
```

(b)
```
    4  2  6
 +  3  8  7
```

(c)
```
    8  6  4
 +  1  3  8
```

(d)
```
    7  7  7
 +  1  4  4
```

(e)
```
    6  0  9
 +  2  5  5
```

(f)
```
    7  5  2
 +  3  4  0
```

(g)
```
    3  8  5
 +  4  9  5
```

(h)
```
    5  0  0
 +  1  0  8
```

3. What number should be added to the given number to make a sum of 50.

(a) 24 + _____ = 50

(b) 38 + _____ = 50

(c) 19 + _____ = 50

(d) 21 + _____ = 50

(e) 35 + _____ = 50

(f) 10 + _____ = 50

(g) 42 + _____ = 50

(h) 1 + _____ = 50

(i) 50 + _____ = 50

(j) 49 + _____ = 50

4. What number should be added to the given number to make a sum of 100.

(a) 75 + _____ = 100

(b) 63 + _____ = 100

(c) 25 + _____ = 100

(d) 14 + _____ = 100

(e) 88 + _____ = 100

(f) 36 + _____ = 100

(g) 90 + _____ = 100

(h) 7 + _____ = 100

(i) 50 + _____ = 100

(j) 42 + _____ = 100

(k) 61 + _____ = 100

(l) 59 + _____ = 100

Subtraction

1. Follow the steps to subtract the numbers and find the answer.

(a)
4	7
− 2	9

(b)
7	8
− 3	5

(c)
8	1
− 5	5

(d)
5	2
− 2	2

(e)
6	5
− 4	0

(f)
9	0
− 8	1

(g)
8	9
− 5	4

(h)
4	0
− 2	8

(i)
7	6
− 2	8

(j)
5	9
− 4	8

(k)
9	3
− 3	9

(l)
8	8
− 6	9

(m)
4	5	6
− 2	4	8

(n)
8	7	7
− 4	5	9

(o)
7	6	9
− 2	8	8

(p)
6	9	3
− 4	2	7

(q)
7	4	5
− 3	6	6

(r)
8	0	0
− 4	3	9

2. Read the following word problems and solve them.

(a)	Kim had $700 in her pocket. She bought a fruit basket for $350. How much money does she have left? Answer: _____	$700 −$350 _____
(b)	There are 864 plants in an orchard, out of which 638 bear flowers. How many plants do not have flowers? Answer: _____	
(c)	1000 students study in a school. If 453 of them are boys, how many girls are there in the school? Answer: _____	
(d)	There were 650 candies in a store in the morning. In the evening, only 204 were left. How many candies were sold that day? Answer: _____	
(e)	There are 145 bees in a hive. If 88 are male bees, how many female bees are there? Answer: _____	
(f)	There are 990 seats in a cinema hall. If 647 seats are occupied, how many more people can sit in the cinema hall? Answer: _____	
(g)	After buying some stationery, Max paid $500 to the shopkeeper. If the shopkeeper returned $25 as change, then how much did the stationery cost? Answer: _____	

Multiplication

1. **Use multiplication tables to solve the following.**

(a) | $7 \times 9 =$ | (b) | $11 \times 6 =$ |

(a) $7 \times 9 =$

(b) $11 \times 6 =$

(c) $9 \times 5 =$

(d) $8 \times 8 =$

(e) $6 \times 12 =$

(f) $4 \times 9 =$

(g) $8 \times 7 =$

(h) $6 \times 7 =$

(i) $10 \times 5 =$

(j) $9 \times 11 =$

(k) $12 \times 8 =$

(l) $8 \times 6 =$

(m) $6 \times 9 =$

(n) $7 \times 7 =$

(o) $9 \times 9 =$

(p) $12 \times 9 =$

2. **Read the statements below and find the answers using multiplication.**

(a) There are 5 groups of bananas with 6 bananas in each group. So there are _____ bananas in all.

(b) There are 9 cookies in one plate. So, 4 such plates will have _____ cookies in all.

(c) There are 3 leaves on one plant and 5 caterpillars on each leaf. So, there are _____ caterpillars in all.

(d) There are **12** beads in a string. Thus, **7** such strings have _____ beads in all.

(e) There are **8** bunches of grapes and **10** grapes in each bunch. Thus, there are _____ grapes in all.

(f) A bird eats **4** worms in a day. It will eat _____ worms in **7** days.

(g) One dog has _____ legs. So, **4** dogs will have _____ legs.

(h) We have _____ fingers on one hand. There are **8** raised hands in class. There are _____ fingers raised in all.

(i) One octopus has _____ tentacles. So, **7** octopuses will have _____ tentacles.

(j) A pickup truck has **6** wheels. Thus, **12** such trucks will have _____ wheels.

3. **Complete the chart by multiplying each number by the number at the centre.**

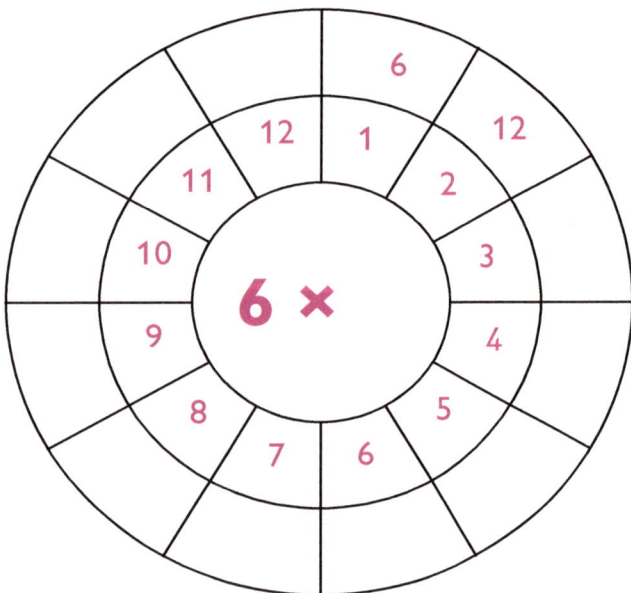

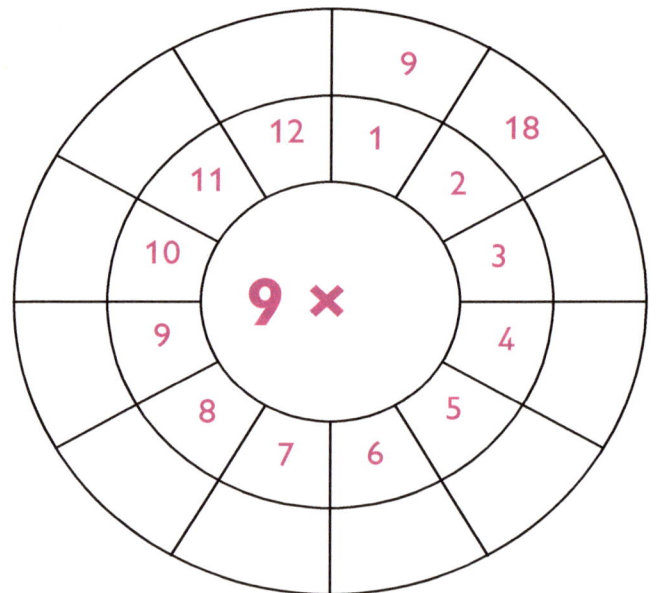

Division

1. **Read the division statements and complete them.**

(a) Make groups of 4 stars by drawing a circle around them.

How many stars are there in all? ☐

How many stars are there in each group? ☐

How many groups of stars will be formed? ☐

20 ÷ 4 = ☐

(b) Make groups of 3 triangles by drawing a circle around them.

How many triangles are there in all? ☐

How many triangles are there in each group? ☐

How many groups of triangles will be formed? ☐

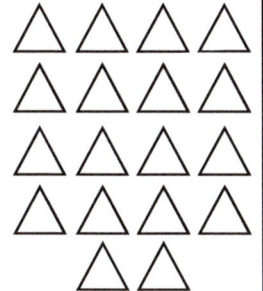

18 ÷ 3 = ☐

(c) Make groups of 6 squares by drawing a circle around them.

How many squares are there in all? ☐

How many squares are there in each group? ☐

How many groups of squares will be formed? ☐

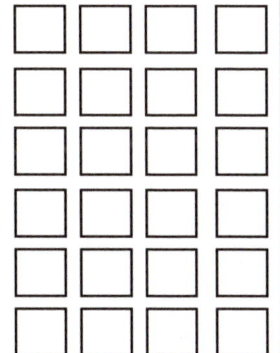

24 ÷ 6 = ☐

2. Divide the following. Write the letter shown with each problem with the answer of the problem given below. Complete the word to get the answer to the riddle. One has been done for you.

A	$45 \div 5 = 9$

U	$36 \div 6 = \_\_\_\_\_$

E	$20 \div 5 = \_\_\_\_\_$

B	$90 \div 9 = \_\_\_\_\_$

L	$27 \div 9 = \_\_\_\_\_$

N	$77 \div 7 = \_\_\_\_\_$

R	$36 \div 3 = \_\_\_\_\_$

A	$36 \div 4 = \_\_\_\_\_$

M	$72 \div 9 = \_\_\_\_\_$

L	$30 \div 10 = \_\_\_\_\_$

Solve the Riddle!

What goes up when rain comes down?

A								
9	11	6	8	10	12	4	3	3	9

Fractions

1. **Shade each shape according to the given fraction.**

(a) 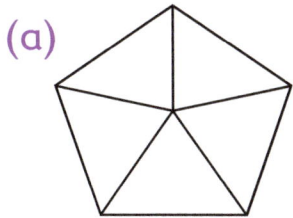 $\dfrac{3}{5}$

(c) 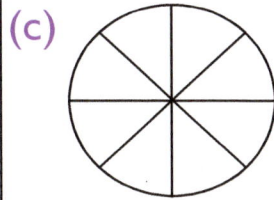 $\dfrac{6}{8}$

(e) 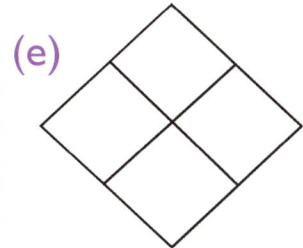 $\dfrac{1}{4}$

(b) 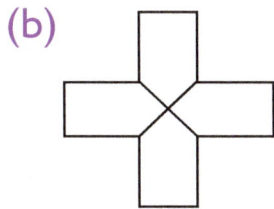 $\dfrac{2}{4}$

(d) 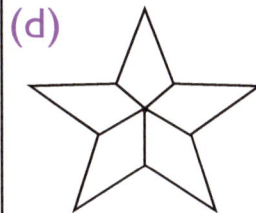 $\dfrac{2}{5}$

(f) 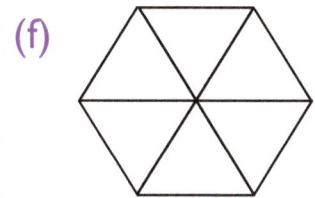 $\dfrac{6}{6}$

2. **Shade one-half of each of the given shapes.**

(a) 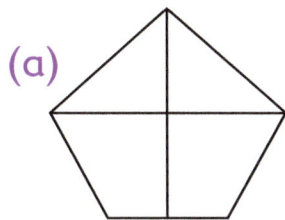 (b) 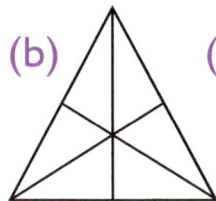 (c) 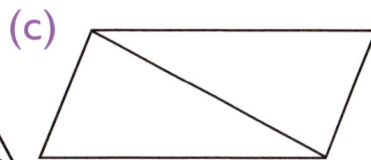 (d)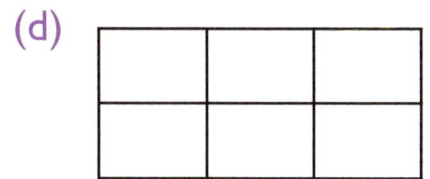

3. **Shade one-quarter of each of the given shapes.**

(a) 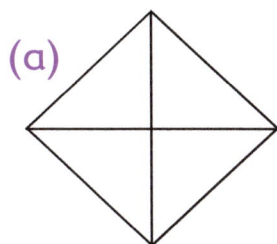 (b) 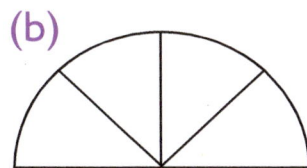 (c) 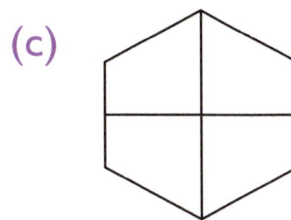 (d)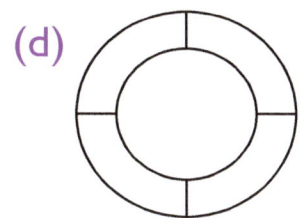

4. **Complete the fraction to show the shaded part of each shape.**

(a) 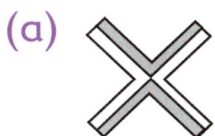 $\dfrac{}{4}$

(b) 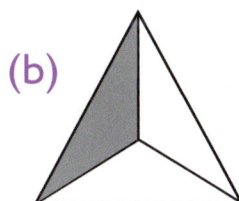 $\dfrac{1}{}$

(c)  $\dfrac{}{5}$

5. Compare the shaded fractions in the shapes using the symbols > , < or =.

(a)

(b)

(c)

(d)

(e)

(f)

6. Draw a circle around the fraction that correctly matches the shaded part of each of the shapes.

(a)

$\dfrac{4}{6}$ $\dfrac{5}{9}$ $\dfrac{4}{9}$

(b)

$\dfrac{2}{3}$ $\dfrac{1}{3}$ $\dfrac{3}{2}$

(c)

$\dfrac{8}{8}$ $\dfrac{4}{8}$ $\dfrac{4}{3}$

24

1. **Break up the numbers to add them. See the example given below.**

(a)

48 + 31

40 + 8 + 30 + 1

40+30 + 8 +1

70 + 9

79

(b)

24 + 53

☐ + ☐ + ☐ + ☐

☐ + ☐

☐ + ☐

☐

(c)

18 + 25

☐ + ☐ + ☐ + ☐

☐ + ☐

☐ + ☐

☐

(d)

63 + 28

☐ + ☐ + ☐ + ☐

☐ + ☐

☐ + ☐

☐

2. **Break up the numbers to subtract them. See the example given below.**

(a)

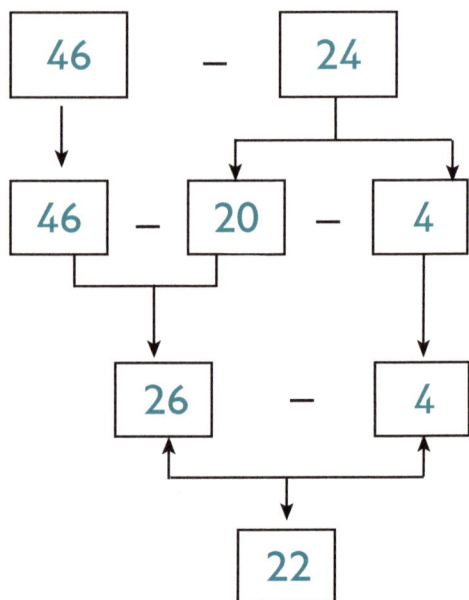

(b)

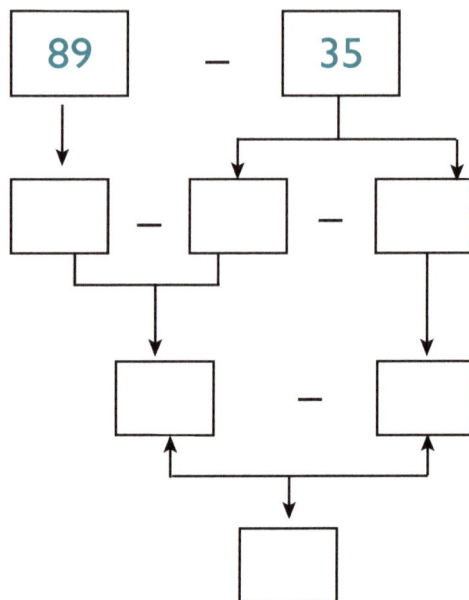

(c)

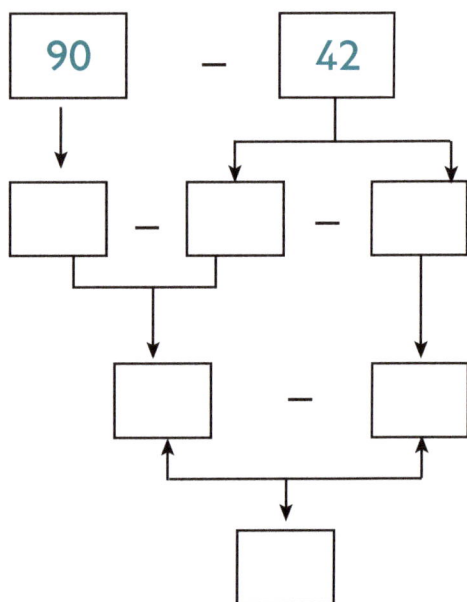

(d)

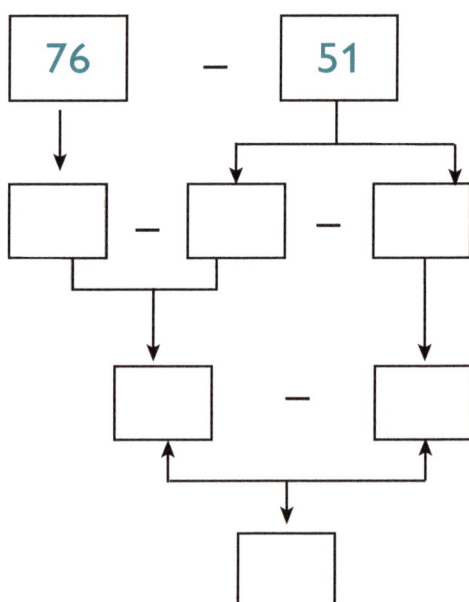

Patterns with Shapes

1. Look at the pattern in each row to fill in the blanks.

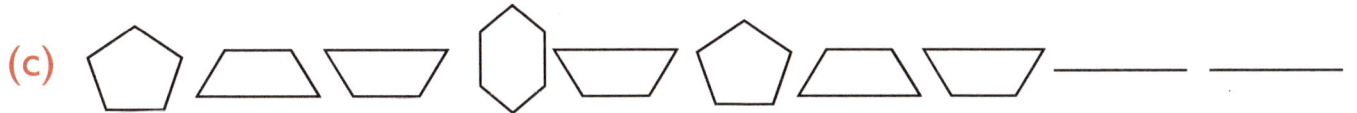

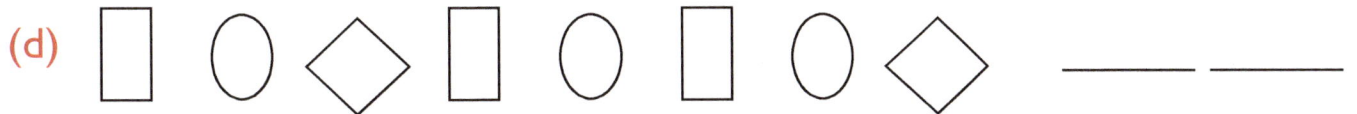

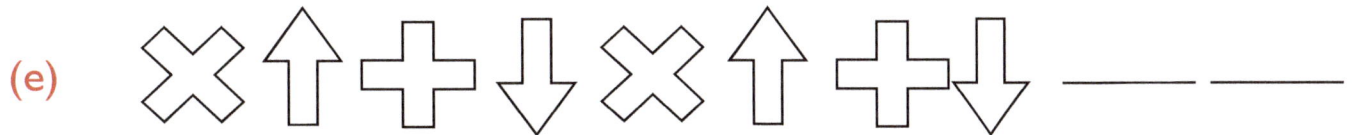

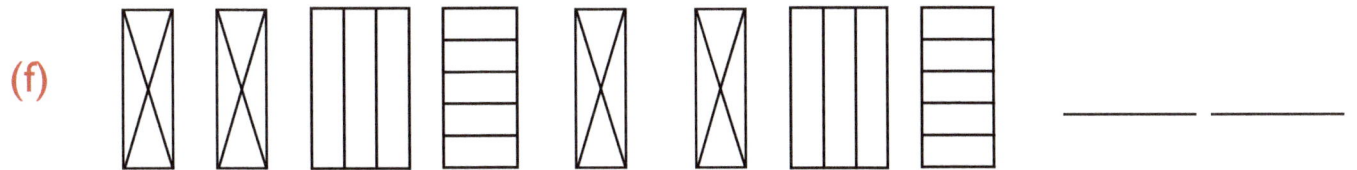

(a) △ ○ ▢ △ △ ○ ○ ▢ △ ____ ____

(b) ✶ ○ ○ ⬠ ✶ ○ ○ ⬠ ____ ____

(c) ⬠ ▱ ▱ ⬡ ▱ ⬠ ▱ ▱ ____ ____

(d) ▯ ○ ◇ ▯ ○ ▯ ○ ◇ ____ ____

(e) ✕ ↑ ✚ ↓ ✕ ↑ ✚ ↓ ____ ____

(f) ▧ ▧ ▥ ▤ ▧ ▧ ▥ ▤ ____ ____

2. Complete the given patterns in the dotted grid.

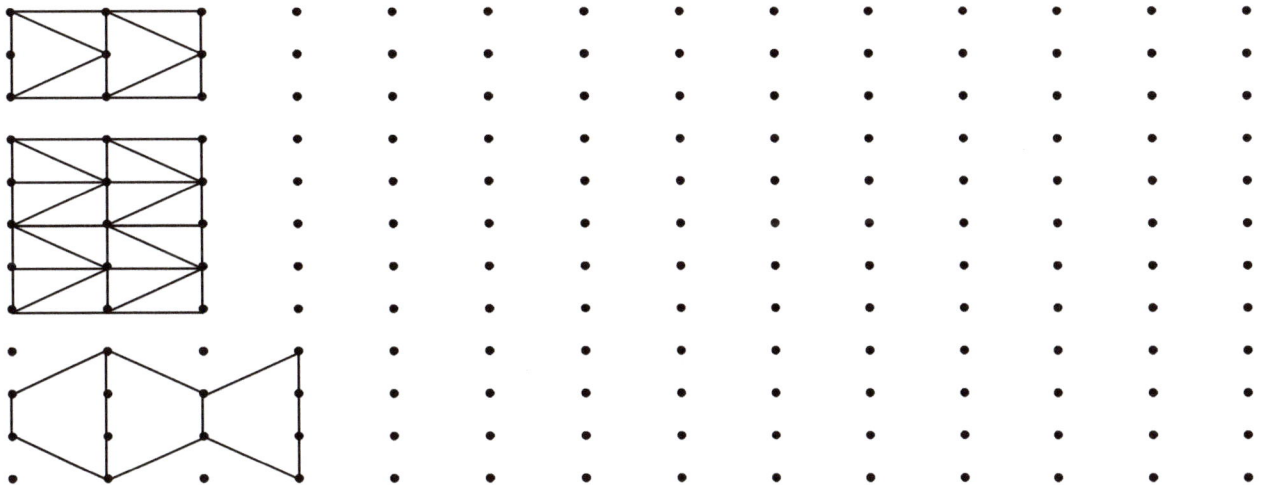

2-D Shapes

1. **Write the names of the shapes below. Take the help of the clue box.**

(a)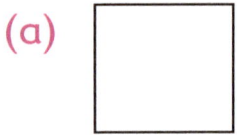

....

(b)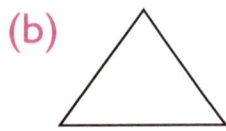

....

(c)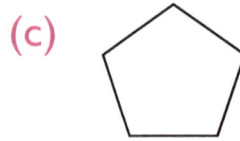

....

(d)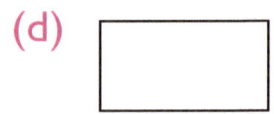

....

(e)

....

(f)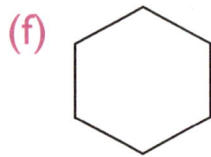

....

(g)

....

(h)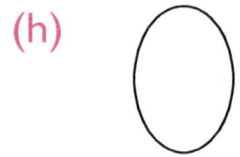

....

(i)

....

(j)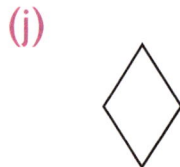

....

Help Box

oval   square   circle   triangle
star   pentagon   diamond
rectangle   heart

2. **Read the names of the objects given below. Join them to the shape that matches their shape.**

Football

Chessboard

Starfish

Honeycomb

Door

Kite

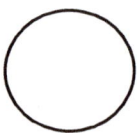

3-D Shapes

1. Name the 3-D shapes given below.

(a) 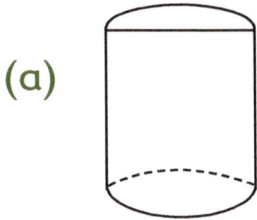   (b) 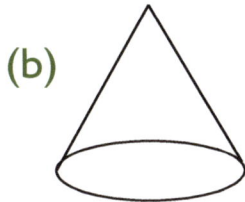   (c) 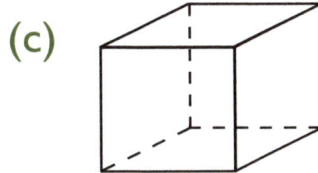   (d)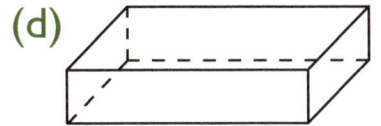

....

2. Name the 2-D shape of the marked face of the 3-D objects below.

(a)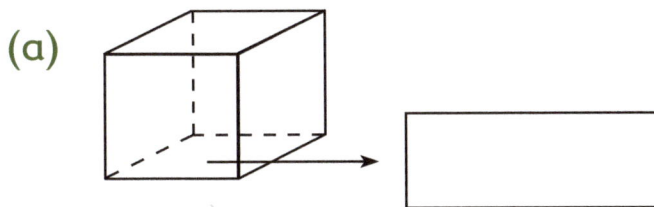

(b)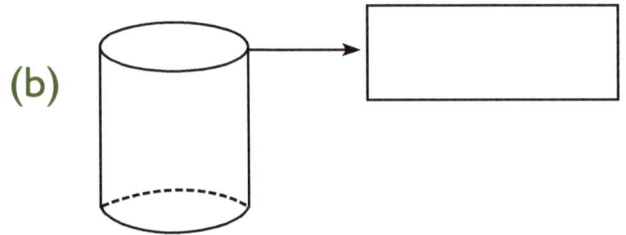

(c)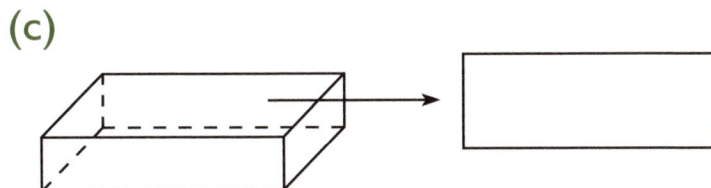

(d)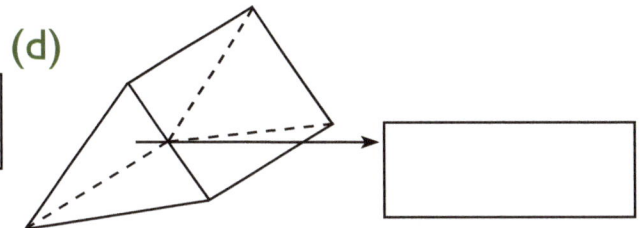

3. Complete the table given below.

3-D shape	Number of faces	Number of vertices	Number of edges
(a) Cube			
(b) Cone			
(c) Cylinder			
(d) Cuboid			

Mirror Images

1. **Draw the mirror images of the given numbers and letters.**

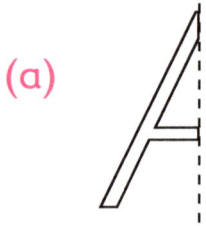

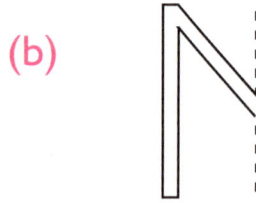

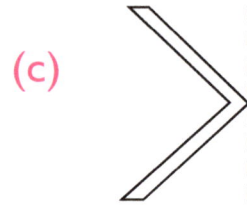

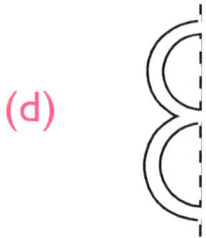

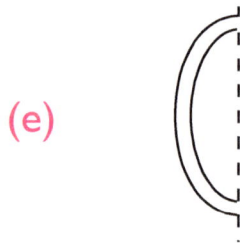

(a)

(b)

(c)

(d)

(e)

(f)

2. **Circle the correct mirror image of the given shapes.**

?			
?			
?			
?			

Money

1. **Solve the problems given below.**

(a) Aman buys eight cupcakes for a party. Each cupcake costs $6. How much does he have to pay in all?

(b) A lady bought some oranges for $35 and some berries for $25. If she gave a $100 note to the vendor, how much change should he return to her?

(c) Lisa bought twelve eggs for $4 each and a can of milk for $12. How much money does she have to pay at the counter?

(d) Two children put up a lemon juice stall. They sold ten glasses of juice for $5 each. If they divide their earning equally; how much money will each kid get?

(e) A milkman sells milk at the rate of $11 per litre. If he sold 9 litres of milk in a day, how much money did he earn?

(f) An old man bought some envelopes for $38 and some postage stamps for $56. How much money does he have to pay in total?

Let's Find the Treasure

Start →	$10	$20	$15	$2

				$3

$13	$9	$7	$25	

How to play

* Throw the dice.

* Read the number on the dice and move your disk by the same number of steps.

• Keep adding the money written on the steps where you land.

• Follow the commands on the step board.

• The player with more money will be the winner of the game.

$15 + $7			You misused money. Go back 6 steps.	You spent more money. Go back two steps.
$3			$2	$4
$30 − $25			$15	$8
Found 2 notes of $15. [+$15×2]			$8	$15
$1			**Treasure**	You saved money. Add $50 to your savings.
				$6
$8	$5	$20 + $10	You lost money on your way. [−$20]	$2 + $8

1. **Colour the given shapes according to the codes given below each shape.**

(a)

$\dfrac{1}{8}$ = Red $\dfrac{2}{8}$ = Blue

$\dfrac{3}{8}$ = Green $\dfrac{2}{8}$ = Yellow

(b)

$\dfrac{1}{6}$ = Pink $\dfrac{2}{6}$ = Violet

$\dfrac{2}{6}$ = Brown $\dfrac{1}{6}$ = Orange

2. **Find the fraction of the shapes which are shaded and those which are not shaded.**

(a)

Shaded Not shaded

(b)

Shaded Not shaded

(c)

Shaded Not shaded

(d)

Shaded Not shaded

3. **Complete the table given below.**

Shape	Number of sides	Number of corners	Number of angles
(a)			
(b)			
(c)			
(d)			

4. **Complete the pattern given below and colour them.**

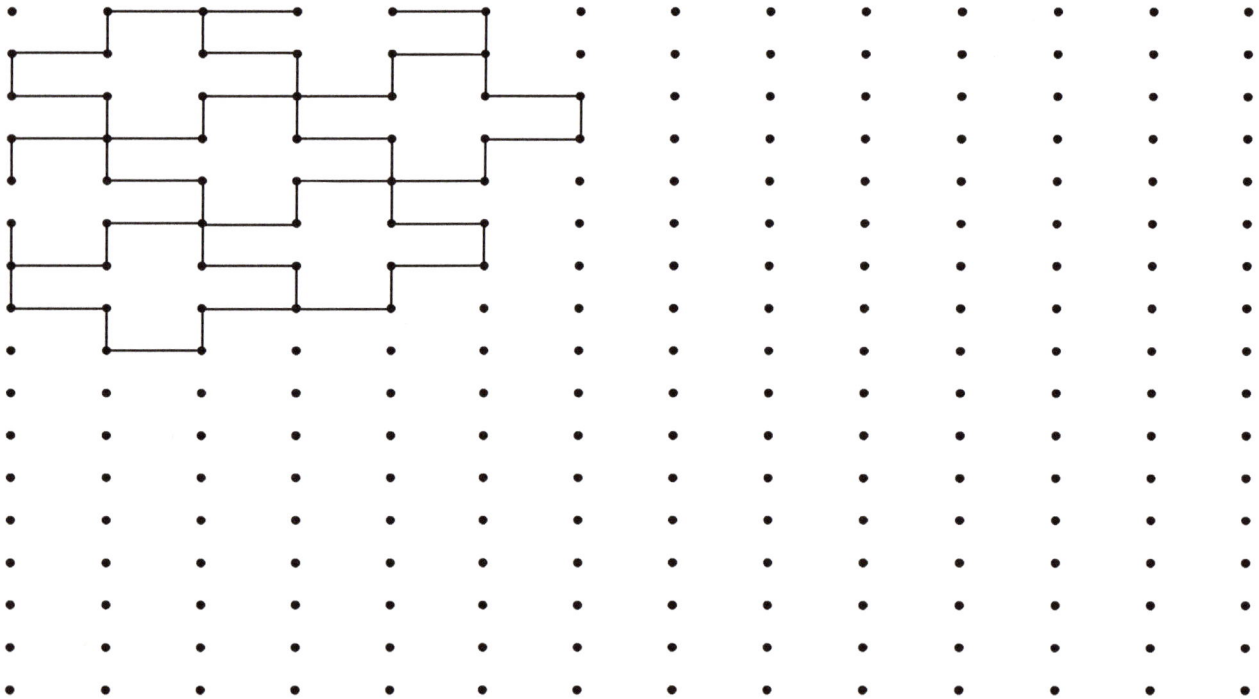

Organizing Data

1. **Fill each type of shape with different colours and then count them to fill the table given below.**

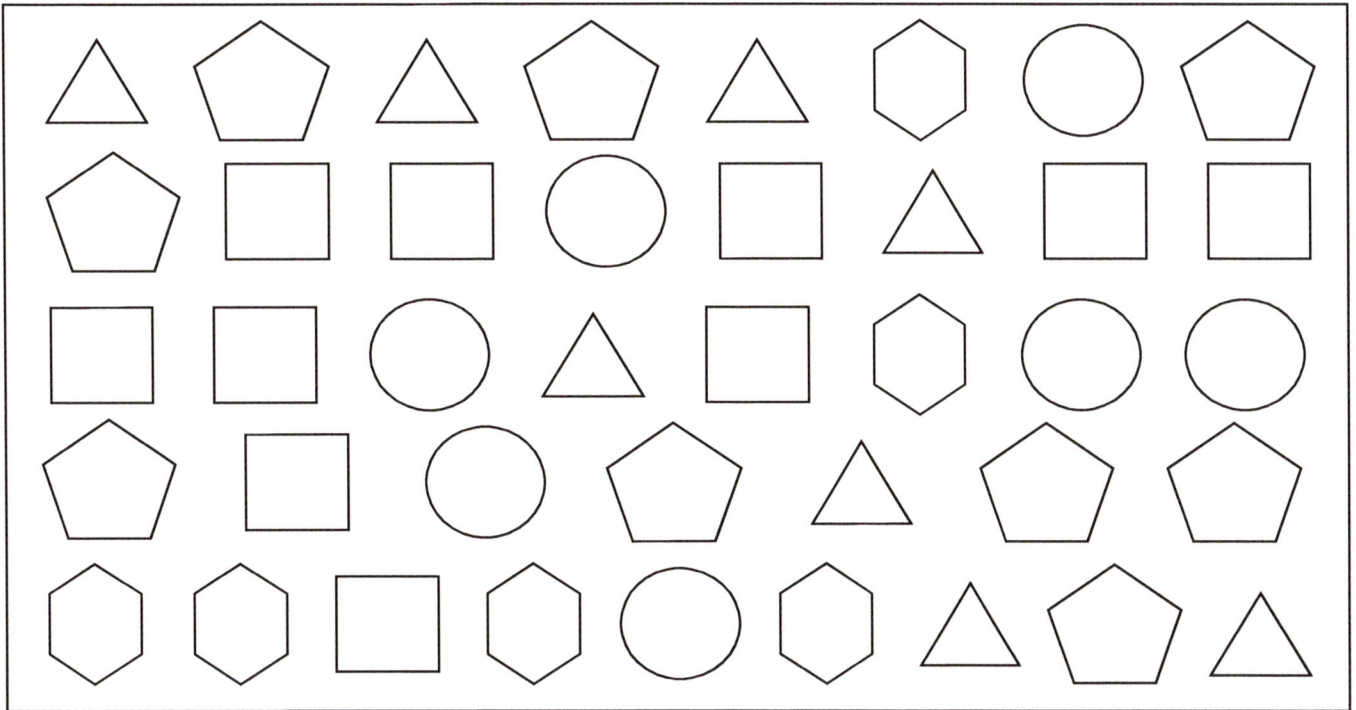

Now let us tally our data.

△	⫲⫲ ⫲ |||	8
⬠		
◯		
☐		
⬡		

2. A fruit-seller keeps a record of his sales in the form of a picture graph. Read the graph and answer the questions below.

Day	Fruits Sold
Monday	🍎 🍎 🍎 🍎 🍎 🍎
Tuesday	🍎 🍎 🍎 🍎 🍎 🍎 🍎
Wednesday	🍎 🍎 🍎 🍎
Thursday	🍎 🍎 🍎
Friday	🍎 🍎 🍎 🍎 🍎 🍎 🍎 🍎
Saturday	🍎 🍎

Key : 🍎 represents 10 fruits.

(a) On which day did he sell the maximum number of fruits? _____

(b) On which day did he sell the least number of fruits? _____

(c) How many fruits did he sell on the first day of the week? _____

(d) How many fruits did he sell on Thursday? _____

(e) How many fruits did he sell on the first two days of the week? _____

Day and Date

1. **Read the calendar for the month of January and write the answers for the questions given below.**

Monday	Tuesday	Wednesday	Thursday	Friday	Saturday	Sunday
1	2	3	4	5	6	7
8	9	10	11	12	13	14
15	16	17	18	19	20	21
22	23	24	25	26	27	28
29	30	31				

(a) What is the day on 18 January? _____

(b) What is the day after 15 January? _____

(c) What is the date on the second Friday of January? _____

(d) What is the date on the last Sunday of January? _____

(e) What is the day on the last date of January? _____

(f) Which are the days that occur maximum number of times in the given calendar? _____

(g) Sara mailed the invitation cards of her birthday party a week before her birthday. If she mailed them on 26 January then her birthday is on _____ .

(h) On which day of the week did Sara have her party? _____

Telling Time

1. **Read the time on each clock and write it in the space given below.**

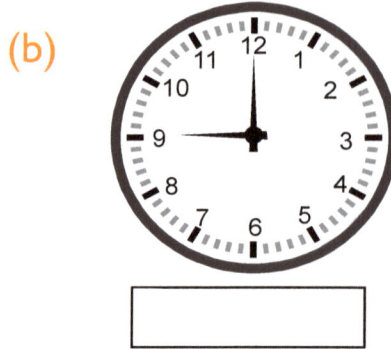

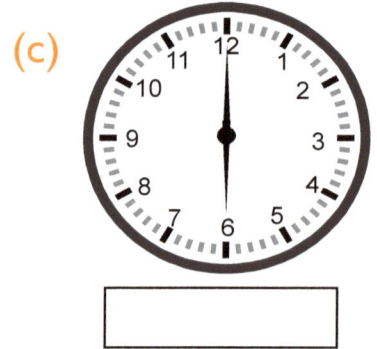

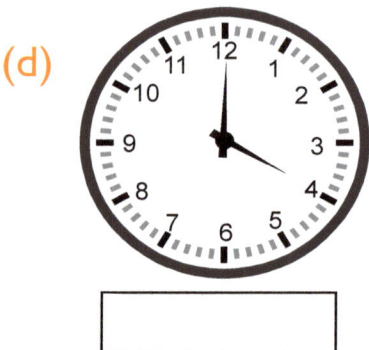

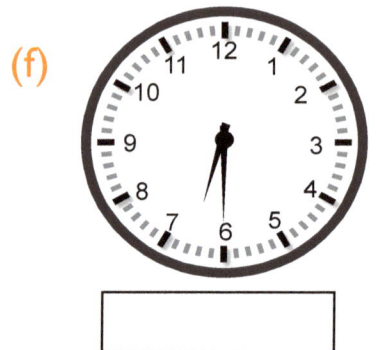

(a)

(b)

(c)

(d)

(e)

(f)

2. **Read the time and draw the hands of the clock.**

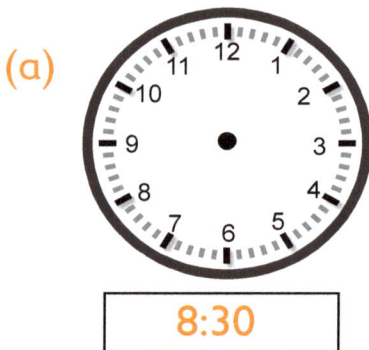

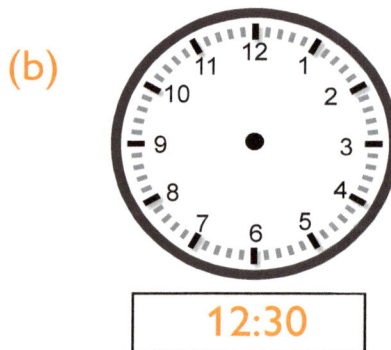

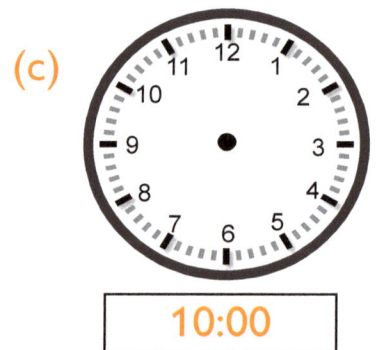

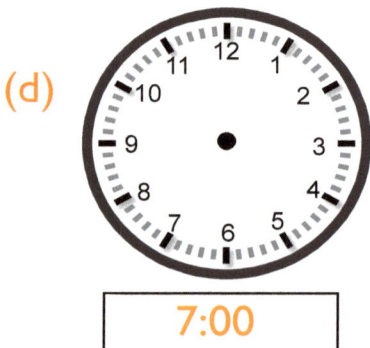

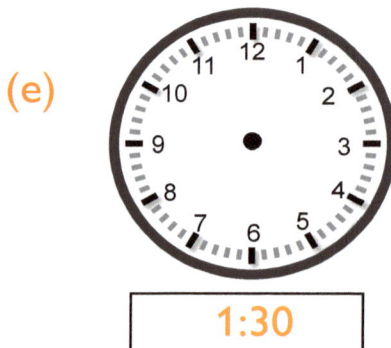

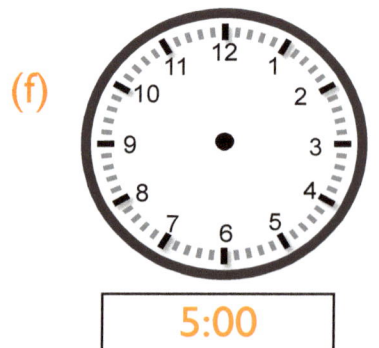

(a)

8:30

(b)

12:30

(c)

10:00

(d)

7:00

(e)

1:30

(f)

5:00

Measuring Length

1. **Measure the lines drawn below using a ruler and write down their lengths. Then answer the questions given below.**

A _____

_____ cm

B _____

_____ cm

C _____

_____ cm

D _____

_____ cm

Answer the following.

(a) Which is the longest line? _____

(b) Which is the shortest line? _____

(c) Which line is longer than line D but shorter than line C? _____

2. **Use a measuring tape to measure the following things. Write their length in metric units.**

(a) The length of your handspan. _____

(b) The length of your pencil. _____

(c) The length of the front door of your house. _____

(d) The length of your mathematics book. _____

(e) The length of your shoe. _____

(f) The length of your water bottle. _____

(g) Your height. _____

(h) Your best friend's height. _____

Measuring Area

1. **Count the number of unit squares in each shape to find out their area.**

(a)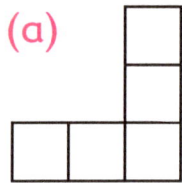

_____ square unit

(b)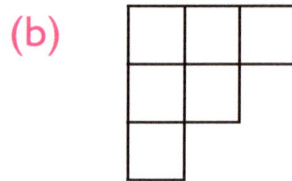

_____ square unit

(c)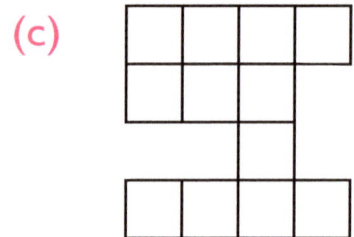

_____ square unit

(d)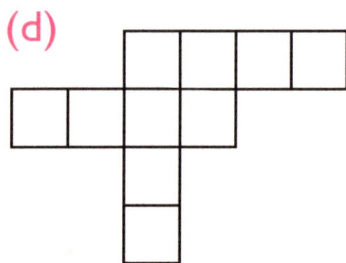

_____ square unit

(e)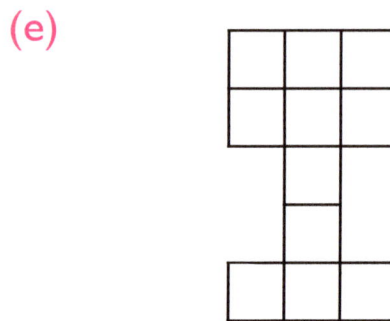

_____ square unit

(f)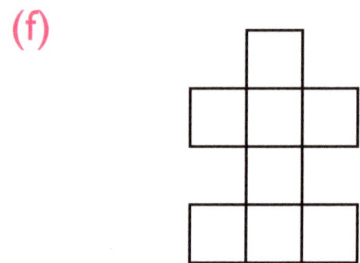

_____ square unit

2. **Use the given square boxes and colour their area according to the codes given below.**

(a) 9 squares (green)   (b) 18 squares (red)

(c) 13 squares (Blue)   (d) 5 squares (yellow)

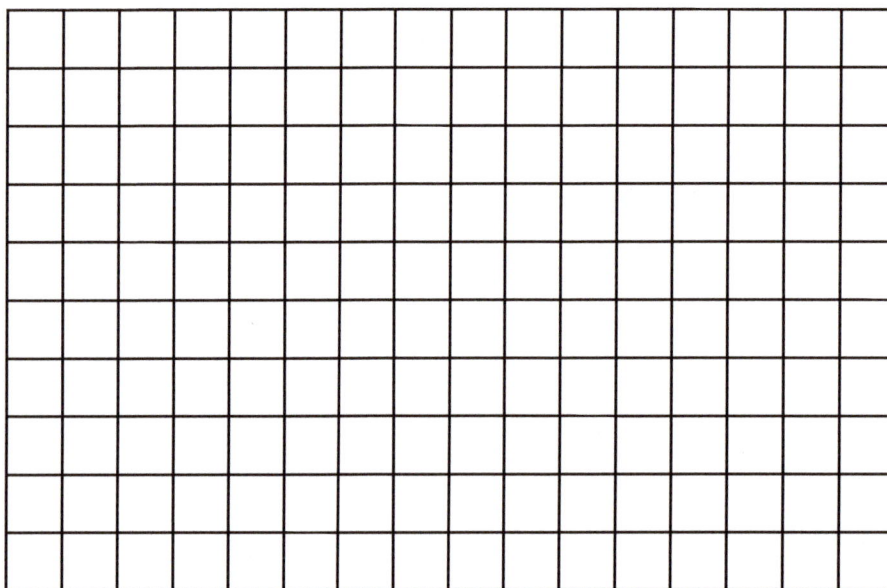

Weighing Things

1. **Circle the object in each group that weighs more than the other.**

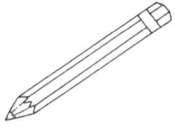

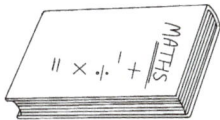

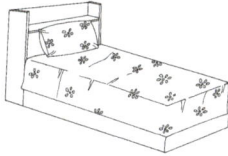

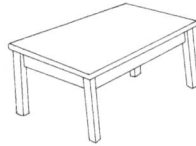

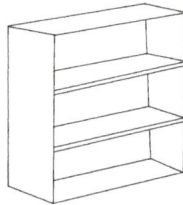

2. **Choose two items in your school bag and measure them. Use a weighing scale to calculate how much each weighs.**

Draw the first object here	Draw the second object here.
Weight = _____	Weight = _____

Test Yourself

1. **Use different number operations to solve the following problems.**

(a) There are twelve chocolates that are divided equally among 3 kids. Thus, each kid will get _____ chocolates.

(b) There are **8** flowers, and there are **2** bees sitting upon each flower. Therefore, there are _____ bees in all.

(c) There are thirty five pieces of bread. If we distribute them equally among 7 dogs, then each dog will get _____ pieces of bread.

(d) There are **24** toys in a room and each child gets one quarter of the total number of toys. Therefore, there are _____ children in the room, and each child gets _____ toys.

(e) There are **38** flowers and **4** vases. If we divide the flowers equally into 4 vases, there will be _____ flowers left out. Each vase will have _____ flowers.

(f) 4 children went to buy candies. One candy costs $2. Each child has different amounts of money. Now, find out how many candies each of the child can buy with the money he or she has.

(i) Child 1 has $8 _____ candies.

(ii) Child 2 has $14 _____ candies.

(iii) Child 3 has $6 _____ candies.

(iv) Child 4 has $20 _____ candies.

2. Answer the following questions. Use a calendar if necessary.

(a) How many days does the month of April have? _____

(b) How many days does the month of August have? _____

(c) How many days are there in February in a leap year? _____

(d) How many Sundays are there in a week? _____

(e) If today is the 7th day of July and it is a Thursday, then what will be:

   (i) the date on the next Thursday? _____

   (ii) the day on the 10 July? _____

   (iii) the day on the first day of the month? _____

   (iv) the day on the last day of the month? _____

3. Solve the following sums and notice how the multiplication and division problems are related to each other.

(a)	7×4 = _____ $28 \div 7$ = _____	(b)	$64 \div 8$ = _____ $8 \times$ _____ = 64
(c)	5×9 = _____ $45 \div$ _____ = 5	(d)	12×4 = _____ $48 \div 4$ = _____
(e)	11×6 = _____ $66 \div 11$ = _____	(f)	10×3 = _____ $30 \div$ _____ = 3

Challenge Yourself

1. Complete the puzzles given below. Look at the solved example to guide you.

(a)
```
    [ 2 ]
( 9 ) + × ( 14 )
    [ 7 ]
```

(b)
```
    [ 6 ]
( ) + × ( )
    [ 4 ]
```

(c)
```
    [ 3 ]
( ) + × ( )
    [ 5 ]
```

(d)
```
    [ 9 ]
( ) + × ( )
    [ 10 ]
```

(e)
```
    [ 5 ]
( ) + × ( )
    [ 12 ]
```

(f)
```
    [ 8 ]
( ) + × ( )
    [ 11 ]
```

(g)
```
    [ 7 ]
( ) + × ( )
    [ 7 ]
```

(h)
```
    [ 11 ]
( ) + × ( )
    [ 3 ]
```

(i)
```
    [ 6 ]
( ) + × ( )
    [ 8 ]
```

(j)
```
    [ 1 ]
( ) + × ( )
    [ 9 ]
```

(k)
```
    [ 2 ]
( ) + × ( )
    [ 2 ]
```

(l)
```
    [ 7 ]
( ) + × ( )
    [ 4 ]
```

(m)
```
    [ 10 ]
( ) + × ( )
    [ 6 ]
```

(n)
```
    [ 5 ]
( ) + × ( )
    [ 5 ]
```

(o)
```
    [ 8 ]
( ) + × ( )
    [ 9 ]
```

Answers

Numbers

1. (a) 244, 245, 246, 247, 249
   (b) 394, 395, 396, 398, 399, 400
   (c) 920, 922, 923, 924, 925
   (d) 837, 838, 839, 840, 841
   (e) 448, 449, 450, 452, 453, 454
   (f) 780, 781, 782, 783, 784, 786
   (g) 158, 160, 161, 162, 163, 164
   (h) 675, 676, 677, 678, 679, 681
   (i) 597, 599, 600, 601, 602
   (j) 532, 533, 534, 535, 536, 538

Comparing Numbers

1. (a) >   (b) >   (c) >
   (d) <   (e) >   (f) <
   (g) >   (h) <   (i) <
   (j) >   (k) <   (l) =
2. (a) 136, 451   (b) 299, 992
   (c) 460, 698   (d) 729, 781
   (e) 178, 875   (f) 100, 600

Ordering Numbers

1. (a) 193, 206, 389, 452, 782
   (b) 991, 883, 647, 527, 400
   (c) 795, 767, 752, 731, 708
   (d) 296, 597, 695, 962, 975
   (e) 793, 739, 397, 379, 97
   (f) 129, 429, 529, 629, 829

Skip Counting

1. (a) 226, 228, 230, 232, 234
   (b) 556, 561, 566, 571, 576
   (c) 869, 872, 875, 878, 881
   (d) 370, 380, 390, 400, 410
   (e) 800, 803, 806, 809, 812
   (f) 973, 978, 983, 988, 993
   (g) 496, 596, 696, 796, 896
   (h) 679, 681, 683, 685, 687
   (i) 500, 505, 510, 515, 520
   (j) 148, 152, 156, 160, 164

Number Names

1. (a) Three hundred forty eight
   (b) Six hundred seventy three
   (c) One hundred ninety two
   (d) Eight hundred eighty five
   (e) Seven hundred fifty one
   (f) Two hundred six
   (g) Four hundred ninety
   (h) Six hundred forty six
   (i) Nine hundred ninety nine
   (j) Five hundred twenty seven
   (k) Three hundred one
   (l) One thousand

2. (a) 87, Eighty seven
   (b) 927, Nine hundred twenty seven
   (c) 819, Eight hundred nineteen
   (d) 239, Two hundred thirty nine
   (e) 22, Twenty two
   (f) 1000, One thousand
   (g) 872, Eight hundred seventy two

Place Value

1. (a) 9 ones   (b) 7 hundreds
   (c) 9 hundreds   (d) 4 tens
   (e) 2 hundreds   (f) 8 ones
   (g) 1 thousands   (h) 7 hundreds
   (i) 0 tens   (j) 4 ones
   (k) 3 tens   (l) 1 hundreds
   (m) 4 hundreds   (n) 6 tens

2. (a) 302   (b) 1000   (c) 888
   (d) 519   (e) 600

3. (a) 1143   (b) 579   (c) 825
   (d) 1052   (e) 438

Expanded Form

1. (a) 897   (b) 1286
   (c) 709   (d) 542
   (e) 360   (f) 931

2. (a) 600 + 40 + 5
   (b) 700 + 90 + 6
   (c) 1000 + 40 + 3
   (d) 500 + 10 + 8
   (e) 400 + 10
   (f) 900 + 7
   (g) 200 + 10 + 2
   (h) 300 + 80

3. (a) 643, 346   (b) 971, 179
   (c) 853, 358   (d) 862, 268

4. (a) >   (b) >   (c) >
   (d) =   (e) <   (f) >
   (g) =   (h) <   (i) <
   (j) >   (k) =   (l) <

Addition

1. (a) 85   (b) 102   (c) 79
   (d) 57   (e) 95   (g) 94

2. (a) 883   (b) 813   (c) 1002
   (d) 921   (e) 864   (f) 1092
   (g) 880   (h) 608

3. (a) 26   (b) 12   (c) 31
   (d) 29   (e) 15   (f) 40
   (g) 8   (h) 49   (i) 0
   (j) 1

4. (a) 25   (b) 37   (c) 75
   (d) 86   (e) 12   (f) 64
   (g) 10   (h) 93   (i) 50
   (j) 58   (k) 39   (l) 41

Subtraction

1. (a) 18   (b) 43   (c) 26
   (d) 30   (e) 25   (f) 9
   (g) 35   (h) 12   (i) 48
   (j) 11   (k) 54   (l) 19
   (m) 208   (n) 418   (o) 481
   (p) 266   (q) 379   (r) 361

2. (a) $350   (b) 226 plants
   (c) 547 girls   (d) 446 candies

(e) 57 bees   (f) 343 people
(g) $475

Multiplication

1. (a) 63   (b) 66   (c) 45
   (d) 64   (e) 72   (f) 36
   (g) 56   (h) 42   (i) 50
   (j) 99   (k) 96   (l) 48
   (m) 54   (n) 49   (o) 81
   (p) 108

2. (a) 30   (b) 36   (c) 15
   (d) 84   (e) 80   (f) 28
   (g) 4, 16   (h) 5, 40   (i) 8, 56
   (j) 72

Division

1. (a) 5   (b) 6   (c) 4
2. An Umbrella

Fractions

1. (a)   (b)
   (c)   (d)
   (e)   (f)

2. (a)   (b)
   (c)   (d)

3. (a)   (b)
   (c)   (d)

4. (a) 2   (b) 3   (c) 9
5. (a) =   (b) >   (c) =
   (d) <   (e) <   (f) >

6. (a) $\frac{4}{9}$ (b) $\frac{1}{3}$ (c) $\frac{4}{8}$

Test Yourself

1. (b) 77   (c) 43   (d) 91
2. (b) 54   (c) 48   (d) 25

Patterns with Shapes

1. (a)   (b)
   (c)   (d)
   (e)   (f)

2-D Shapes

1. (a) square   (b) triangle
   (c) pentagon   (d) rectangle
   (e) circle   (f) hexagon
   (g) star   (h) oval
   (i) heart   (i) diamand

3-D Shapes

1. (a) cylinder   (b) cone
   (c) cube   (d) cuboid

2. (a) Square   (b) circle
   (c) rectangle   (d) triangle

3. (a) 6, 8, 12   (b) 2, 1, 1
   (c) 3, 0, 2   (d) 6, 8, 12

Money

1. (a) $48   (b) $40   (c) $60
   (d) $25   (e) $99   (f) $94

Oraganizing Data

1. (a) Friday   (b) Saturday
   (c) Sixty   (d) Thirty

(e) One hundred thirty

Day and Date

1. (a) Thursday
   (b) Tuesday
   (c) 12 January
   (d) 28 January
   (e) Wednesday
   (f) Monday, Tuesday, Wednesday
   (g) 2 February
   (h) Friday

Telling Time

1. (a) 2:30   (b) 9:00   (c) 6:00
   (d) 4:00   (e) 1:00   (f) 6:30

Measuring Length

1. (a) Line C   (b) Line A
   (c) Line B

Measuring Area

1. (a) 5   (b) 6   (c) 12
   (d) 10   (e) 11   (f) 8

Test Yourself

1. (a) 4   (b) 16
   (c) 5   (d) 4, 6
   (e) 2, 9   (f) (i) 4
              (ii) 7
              (iii) 3
              (iv) 10

2. (a) 30 days   (b) 31 days
   (c) 29 days   (d) 1
   (e) (i) 14 July
       (ii) Sunday
       (iii) Friday
       (iv) Sunday

Challenge Yourself

1.

(a)
```
        2
   9  +   ×  14
        7
```

(b)
```
        6
  10  +   ×  24
        4
```

(c)
```
        3
   8  +   ×  15
        5
```

(d)
```
        9
  19  +   ×  90
       10
```

(e)
```
        5
  17  +   ×  60
       12
```

(f)
```
        8
  19  +   ×  88
       11
```

(g)
```
        7
  14  +   ×  49
        7
```

(h)
```
       11
  14  +   ×  33
        3
```

(i)
```
        6
  14  +   ×  48
        8
```

(j)
```
        1
  10  +   ×   9
        9
```

(k)
```
        2
   4  +   ×   4
        2
```

(l)
```
        7
  11  +   ×  28
        4
```

(m)
```
       10
  16  +   ×  60
        6
```

(n)
```
        5
  10  +   ×  25
        5
```

(o)
```
        8
  17  +   ×  72
        9
```

www.ingramcontent.com/pod-product-compliance
Lightning Source LLC
Chambersburg PA
CBHW041451210326
41599CB00004B/218